UNDERSTANDING E=mc²

A clear and concise explanation of Special Relativity

Tony Reynolds

Copyright © 2025 Tony Reynolds

All rights reserved. No part of this publication may be reproduced, stored in a retrieval system, or transmitted in any form or by any means—electronic, mechanical, photocopying, recording, or otherwise—without the prior written permission of the author, except for brief quotations used in reviews, academic works, or articles, with appropriate citation.

Every effort has been made to ensure the accuracy of the information presented in this book. However, the author and publisher make no representations or warranties regarding the completeness, accuracy, or applicability of the content, and disclaim any liability for errors or omissions. The information is provided for general educational purposes and does not constitute professional advice.

First edition
ISBN: 979-8-2884685-8-2

Illustrations and cover design by Natia Gogiashvili.
Printed in UK.

The author welcomes comments and suggestions.
Please contact **publish@arcenciel.com**

To –
The memory of H. Rex Harvey, an inspirational physics teacher.

Acknowledgement:
I am grateful to Dr. Ali Azmi for his helpful comments and advice on the text and derivations.

TABLE OF CONTENTS

1. **It's not that difficult!** — 1
   Introduction

2. **The speed of light** — 2
   The constant speed of light

3. **The speed of time** — 5
   The observed passing of time

4. **Suppose we meet up?** — 7
   The twin paradox

5. **We're getting squashed!** — 9
   Contraction in the direction of motion

6. **There is no "Now"** — 11
   Simultaneity

7. **Mass is another form of energy!** — 12
   Equivalence of mass and energy

8. **There's a limit!** — 14
   Light speed as a limit

9. **It was a team effort!** — 16
   Precursors to Einstein

10. **But does it matter to me?** — 18
    Relativity in everyday life

11. **The Maths** — 20
    Mathematical proofs

IT'S NOT THAT DIFFICULT!

Many people when they hear references to the equation $E=mc^2$ or to the "theory of relativity" simply shrug and assume that only scientists can actually understand what they mean.

In fact, it's not particularly difficult! I believe that after you have studied these few pages you will understand how it is that time can pass at different rates in different places and how it is that energy and mass are interchangeable.

To be more precise, there are two theories of relativity each applying to a different situation.

Albert Einstein published articles in 1905 dealing with the case of an observer watching something moving past at a constant speed, as for example when you see a train go by. In this situation Einstein showed that time on the train will appear to you to move more slowly, the length of the train will appear to shorten, and that mass and energy are two forms of the same thing. This is what we now call the "Special Theory of Relativity".

Einstein continued his work, and ten years later solved the more complicated situation of an observer watching something that is accelerating or decelerating, rather than moving at a constant speed. This is known as the "General Theory of Relativity" and describes how massive objects warp space and time.

This book will concentrate on the implications of the Special Theory.

THE SPEED OF LIGHT

Light travels extremely fast – we know now that it moves at over 186,000 miles per second – and it was only relatively recently that scientists acquired the tools to measure it.

You can think of light as being made up of really small particles streaming out of a source such as the sun or perhaps a light bulb. These particles are called photons. In a way they're like very tiny tennis balls. They are emitted from a light source, bounce off objects and then into your eye: that's how you see things.

BUT... there's one way in which photons are not at all like tennis balls, and it's this difference that is key to understanding relativity.

Suppose you're playing a game of tennis and your opponent hits the ball towards you at 20 mph. To return it, you need to run forward at 10 mph, so the ball hits your racquet at a combined speed of 30 mph.

You might think that photons behave in the same way. We race around the sun at about 67,000 mph and the earth spins at about 1000 mph at the equator so you'd assume that as the day goes by, photons from the sun would be reaching your eyes at different speeds. But they don't! YOU ALWAYS SEE THEM MOVING AT EXACTLY THE SAME SPEED!!

This doesn't seem to make sense, but has been proved to be true in many ways, both experimental and theoretical. If you can hold on to this fact, you will understand why the principles of relativity must follow, and in a page or two, you'll see how it comes about.

James Bradley

The first person to find an accurate value for the speed of light was the Astronomer Royal, James Bradley, in 1725. Finding that he had to tilt his telescope slightly away from the expected position of a star, he realised that because the earth was moving, the tilt was needed to allow photons to pass down the telescope.

He calculated the speed of light to be 10,210 times the speed of the earth. Using the best estimate of the earth's speed at the time gave the speed of light as 190,000 miles per second – only about 2% above the true value.

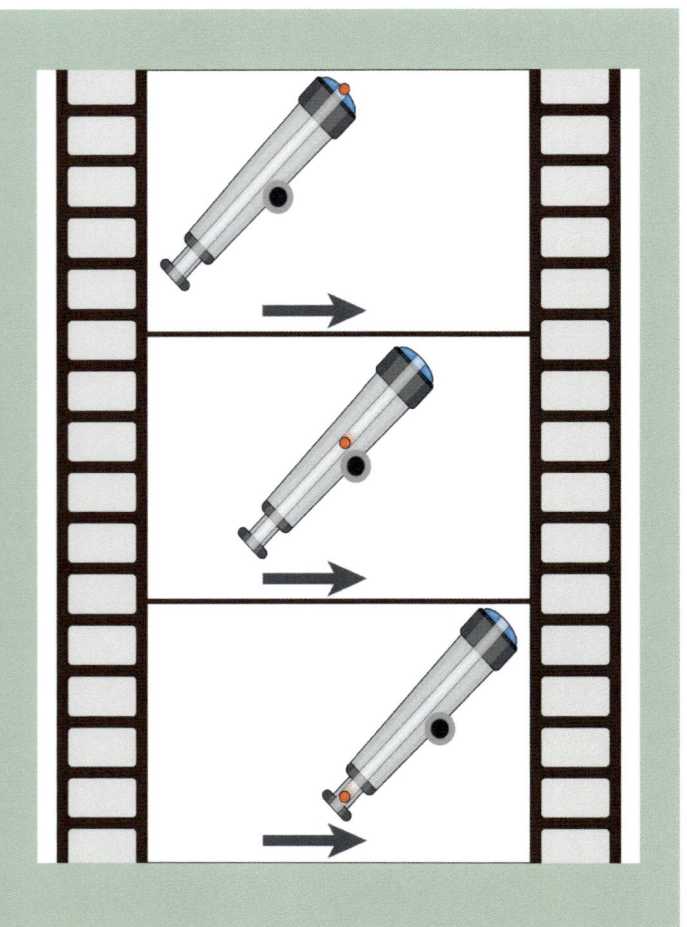

We've heard that the speed of light is always the same everywhere, but if you measure its speed while passing through a medium such as air or water you will observe a lower value. This is because the photons bounce from one molecule to another as they travel, as in the diagram. So because the length of the path is longer, the time taken through the medium is longer. The speed of the photons themselves has not changed.

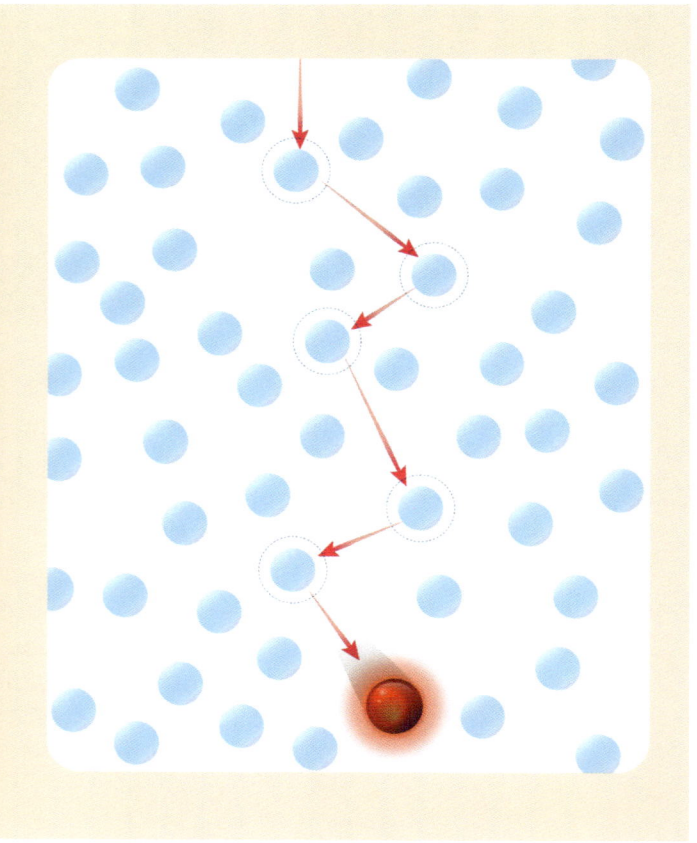

THE SPEED OF TIME

It was Einstein who worked it out by simply saying: OK, let's accept that light *does* always travel at the same speed. What follows from that?

Imagine that you are standing on the platform of a country station just as night is falling and an express train speeds past...

As you watch, you see a passenger switching on the overhead light in their carriage.

Now we have two points of view; to those on the train, the photons are travelling straight down at the speed of light.

But you, the observer, have a different view. To you, each photon on its journey from the light to the floor traces a diagonal line, because the train is moving relative to your position.

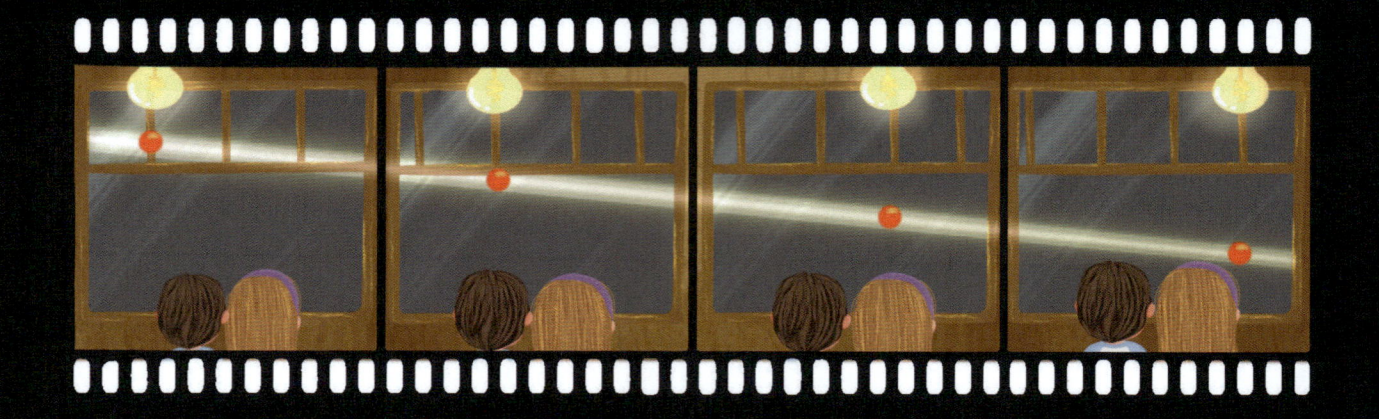

So the passengers see the photons travel a shorter distance than you do.

That's no problem you might think, because the photons have been given a boost by the speed of the train and travel faster along the diagonal, so everything works out for both observers.

BUT WAIT! We've just heard that it's been proved that light is always observed travelling at the same speed. So how can this be? The only conclusion you can reach is that *from your point of view time is moving more slowly on the train!*

The faster the train is moving, the longer the diagonal, and therefore the slower time appears to flow.

This is the first result of the theory of Special Relativity and is often expressed as: "moving clocks run slower". Put in more formal language, you are in one "frame of reference" and the train, which is moving at a constant speed relative to you, is in another frame.

Now we've understood this, the idea of the speed of light being constant for everyone isn't so weird. The definition of speed is something moving a certain distance *in a certain time:* feet per second, kilometres per hour and so on. If there is no such thing as a universal time, then observed speeds will depend on your frame of reference.

SUPPOSE WE MEET UP?

We've spoken about how the observer on the platform sees matters, but what about the passengers on the train?

"Everything is relative", as they say, so they could consider that they were standing still and that the station, and indeed the whole world, was rushing past them.

So everything we talked about before applies to the view outside the train. The passengers might see a station clock and conclude that it was running slow.

This seems very strange: how can the boy see the station clock running slow while someone on the platform sees the boy's watch losing time? Who is right? If you stop the train and let everybody meet up and compare watches, what would they see?

But to meet up, the train would have to slow down – decelerate. It would no longer be moving at the constant speed which is the basic premise of Special Relativity.

Albert Einstein

"Time is an illusion."

In writing to the family of a deceased friend Einstein said: *"Now he has departed from this strange world a little ahead of me. That means nothing. People like us, who believe in physics, know that the distinction between past, present, and future is only a stubbornly persistent illusion."*

The situation where one frame of reference is accelerating or decelerating relative to another is far more complex. It took Einstein ten years and a mountain of advanced mathematics to work out what would happen. The culmination was the General Theory of Relativity, a monumental achievement that melds time and space into a single, unified entity called spacetime. Rather than explaining gravity as a force, Einstein showed that massive objects curve the fabric of spacetime itself, and this curvature guides the motion of objects, which we perceive as the effect of gravity.

The General Theory is for another time. Suffice to say that it predicts that if the train should stop, the passengers on the train will have aged less than those on the platform. This is often referred to as the 'Twin Paradox', but it's not really a paradox, but rather a consequence of the predictions of relativity.

WE'RE GETTING SQUASHED!

Thinking further about the case of a passenger on a train switching on the light, you might come up another way of arranging a light beam.

Instead of turning on the overhead light, let's say that you have two children sitting opposite each other. The boy, who is facing the direction of travel, has a hand torch and the girl has a mirror. They play a game where the boy shines his torch on the mirror as below. So now photons from the light source travel horizontally to the mirror and bounce back.

Now from outside the train, you see the light coming from the torch taking longer to reach the mirror, because the mirror is moving forward and the speed of light is independent of the speed of the train.

BUT the reflected light takes LESS time to return to the starting point because the torch is moving to meet it.

So in this experiment, where all we've done is change the path the photons take, no difference in timing between inside and outside the train would be observed. How can this be reconciled?

The answer forced on us is the second result of the theory of Special Relativity: *objects moving relative to you appear to contract in the direction of motion.* To us, the carriage has become shorter, so the total distance the light has to travel has become less.

The Terrell-Penrose effect

A popular-science book first published in 1940 described how the world would look if the speed of light were no more than 10 mph. In it, the protagonist sees a passing cyclist appearing squashed up, as illustrated below:

Following research by the physicists Roger Penrose and James Terrell, the consensus today is that because of the differing time light rays take to reach an observer, a passing object would appear to be rotated rather than contracted.

THERE IS NO 'NOW'

Let's try another experiment. This time both the boy and the girl have hand torches. Their carriage is in the exact middle of the train. As the train passes through a station, they both lean out of a window (this is dangerous, don't do it for real!) and the girl shines her torch towards the guard's van at the rear of the train and the boy towards the engine at the front.

You are on the platform and you see them do this. Now again we have two points of view: to the passengers on the train, each light beam reaches either end of the train simultaneously, as the distances are the same. But you don't agree with this. As the speed of light is not affected by the speed of the train, you see the girl's beam arrive at the guard's van before the boy's beam reaches the engine because the train is moving forward.

The length of the train is affected by relativistic contraction, but as this applies to both the front and the back halves equally, we can ignore this.

This is the next result of the theory of Special Relativity: *events that are simultaneous in one frame of reference may not be simultaneous when seen from another.*

MASS IS ANOTHER FORM OF ENERGY!

There's one more, and very important, conclusion that we can draw from time slowing down.

Let's imagine that our children are sitting side by side in a railway carriage (side by side so we don't have to worry about contraction in the direction of travel)

They're playing a game by tossing a tennis ball to each other.

Now the energy in a moving object depends on its mass (weight) and the speed at which it travels.

A thrown cricket ball, for example, represents more energy than a tennis ball because it's much heavier. On the other hand, a bullet fired from a gun is very light but has lots of energy because of the speed at which it travels.

You are observing this game from the platform as the train passes and you note that because of time slowing down on the train the ball is moving more slowly than you would expect.

So what's happened here? Is there somehow less energy in the ball as it's travelling more slowly? No, there can't be: it is universally accepted that energy can be neither created nor destroyed, only changed into another form.

So if the energy is made up of mass and speed and the speed decreases, we're forced to the conclusion that to us, the mass of the ball has increased. The ball appears heavier to us watching from the platform than it does to the children on the train.

This is the culmination of the Special Theory – *mass is another form of energy.*

The maths is fairly advanced, although still within the grasp of a high-school student in the science cohort, and with it you can derive the famous relationship: $E = mc^2$, where c is the speed of light. There's a mathematical proof of this in the last chapter.

The Atom

It's a bit easier for us in modern times to understand that mass and energy are different aspects of the same thing than it was in 1905 when Einstein published the Special Theory.

At that time, they visualised an atom as being made up of particles called electrons orbiting a nucleus, but since quantum theory was developed, we know that electrons aren't really particles. It's more correct to regard them as a wave with no exact position, only a shell around the nucleus of greater or lesser probability of being at a given point. This means that electrons can "tunnel" through solid barriers. So at the most fundamental level, matter and energy are different ways of looking at the same thing.

THERE'S A LIMIT!

In day-to-day use we don't notice these relativistic effects because they're so small. If a jet fighter flashes above you at a 1500 mph you'd only see it contract by about 0.000000002 inches. It's only when you get near the speed of light that relativistic changes become noticeable.

But suppose our train was travelling at the speed of light – what then?

In our first experiment, when we observed the switching on of an overhead light, we saw that as the train moves faster and faster the photons appear to you to trace a longer path to the floor, from which we concluded that time was moving slower and slower. At the speed of light, the photons could not escape the bulb – time on the train will have stopped!

In the second experiment, where the children bounced a light off a mirror, the beam would never reach the mirror because the mirror was moving away at the same speed the photons travel. As the speed increased, the train would have to contract more and more to keep the time of travel of the torch beam constant. For light speed to be attained, the length of the train would have to become zero.

In the third experiment, where the children tossed a ball to each other, as time has stopped, the ball will not move and its mass will appear to be infinite.

So we come to the conclusion: *the speed of light is not only a universal constant but also the upper limit at which anything can move.*

But can't you get round it, you might say? Let's imagine that one day we can build a spaceship that travels at three-quarters the speed of light. At some point, it fires a torpedo in the forward direction at half the speed of light. Wouldn't you then have exceeded the light-speed limit?

Well, here again we have two points of view. Inside the spaceship, the captain gives the order to his torpedo officer, who salutes smartly and quickly goes to his position in the bow to fire the missile. The captain observes the earth receding behind him at three-quarters the speed of light and the torpedo racing ahead of him at half the speed of light.

BUT if you on earth could see inside the spaceship at that time, you would see that everybody was moving rather slowly. Because of the relativistic slowing of time, for every second that passes on earth, less than three-quarters of a second would pass in the spaceship. The captain would drawl out his orders and the officer would saunter to his post to fire the missile which would exit the torpedo tube at a comparatively leisurely pace.

The upshot is that if you measured the speed of the torpedo relative to the earth, you would see it travelling at only nine-tenths of the speed of light.

So no-one, in either frame of reference, sees anything exceeding the speed of light.

IT WAS A TEAM EFFORT!

The work of Einstein is so famous that to many people his achievements overshadow those of the scientists that laid the foundations. But as Newton once said, "If I have seen further, it is by standing on the shoulders of giants." and it is only right to acknowledge the contribution of the early workers in the field.

Einstein himself certainly did. We know that he kept pictures of Newton, Faraday and James Clerk Maxwell on his study wall at Princeton. You can learn more about these people below.

Isaac Newton

It could be said that science as we know it began with Isaac Newton, who was born in the time of Charles I. Newton was the principal inventor of the calculus – a form of higher mathematics – and using it derived the laws of gravity and of motion.

Einstein's work almost 300 years later showed that for bodies moving at velocities approaching the speed of light, Newton's results were no longer accurate, but they are still used for most practical purposes.

Michael Faraday

Faraday's experiments in pre-Victorian times laid the foundations for electronics. He came from a poor family and received only the most basic education. Consequently, he had no knowledge of mathematics beyond elementary arithmetic.

Despite this, his brilliant conceptual thinking combined with great experimental skills enabled him to make basic discoveries in electricity and magnetism.

It was already known that an electric current flowing through a wire would deflect a compass. Faraday had the idea that the reverse might also be true – a magnetic field would cause an electric current to flow in a wire.

Faraday tested this by wrapping a paper cylinder with wire to make a coil. He connected the coil to a galvanometer then introduced a magnet into the cylinder. He found that a current did indeed flow, but only as he moved the magnet back and forth.

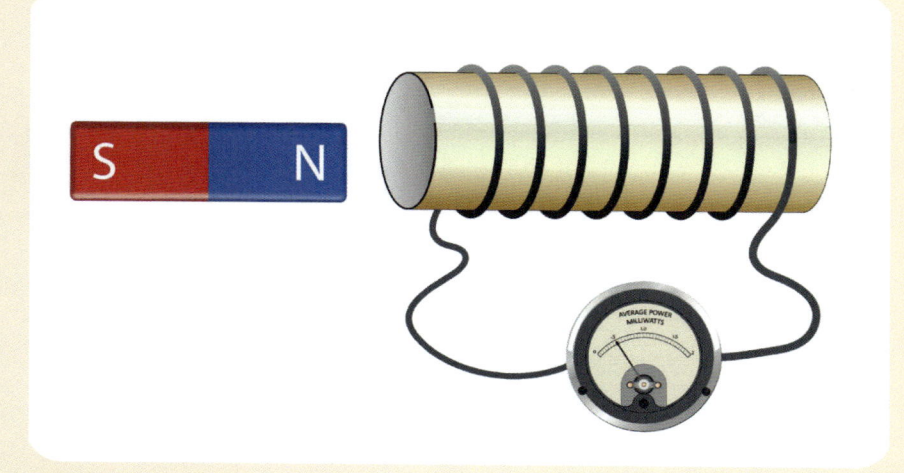

This is the working principle of the dynamo, although in a practical device the magnet is kept stationary and the coil moves.

With this and many other experiments he showed that electricity and magnetism were closely connected.

James Clerk Maxwell

Unlike Faraday, the Scottish researcher James Clerk Maxwell had a strong mathematical background and was able to take Faraday's experimental findings of thirty years previously and provide a rigorous and quantitative framework for them.

He derived a set of four fundamental equations, two of which described mathematically how a changing electric field would produce a magnetic field and vice versa.

Taking this a step further, he reasoned that this meant that a wave could travel through space as an electric field generated a magnetic field which in turn generated another electric field and so on.

Using his equations to find the speed of this wave in a vacuum he arrived at a figure that was very close to the known speed of light. Maxwell did not believe that this could be a coincidence and in a paper he published in 1865 penned the grandiose sentence: "We can scarcely avoid the conclusion that light consists in the transverse undulations of the same medium which is the cause of electric and magnetic phenomena."

Importantly, Maxwell's equations did not include a term for the initial speed of the wave, implying that the velocity of light was always the same, even if the light source were moving relative to the observer.

This is the strange result we encountered at the start of this book, and by the dawn of the twentieth century was one of the great puzzles of physics. It didn't seem to make sense! It was this problem that Einstein set himself to solve.

BUT DOES IT MATTER TO ME?

You might think that because relativistic effects occur in such extreme circumstances they don't actually have much bearing on everyday life.

Well, actually they do. For a start, the sun wouldn't work, which would have a serious impact on your existence.

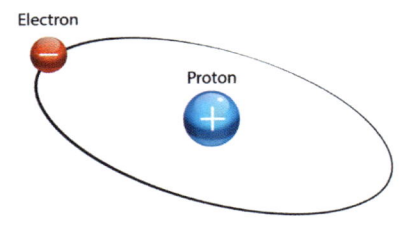

The sun is a huge ball of hydrogen, so big that its heat is continually ripping apart the hydrogen atoms, which have one proton and one electron, and crushing them into helium atoms which have two of each (as well as a couple of neutrons which somehow appear during the fusion process).

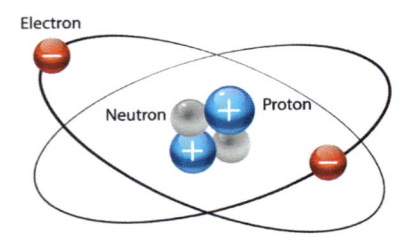

Strangely, a single helium atom weighs a little less than two hydrogen atoms and as we have shown the equivalence of mass and energy, the loss in mass is manifest in the heat and light of the sun.

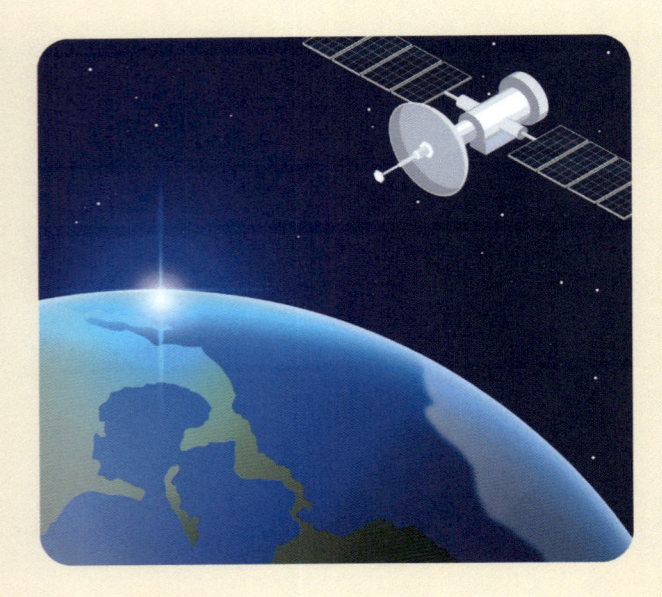

While perhaps not so critical, the relativistic slowing of time is important to modern life. At present, there are thousands of satellites orbiting the earth. They perform important functions such as weather monitoring, global communications and navigation. All of these satellites need to know which points on the earth they are travelling over. The only way to do this is to have an extremely accurate clock on board, ticking at typically 10 million times per second. Depending on the altitude of the satellite, a fixed distance on earth is covered for each tick. But as we've seen, "moving clocks run slower" and as the satellites are racing around at perhaps 10,000 mph and the clocks are so very accurate, relativity has a significant effect.

In this case, although Special Relativity makes the clocks appear to run slower by 7 millionths of a second per day, the effect of General Relativity is much larger. In his later work, Einstein showed that the presence of mass also slows time, so as the satellites are well above the earth, in aggregate the clocks appear to us to run *faster* by about 45 millionths of a second per day.

So if relativistic time changes were not taken into account, satnav wouldn't work and we would all be struggling with paper maps.

Another major effect of Special Relativity that we've seen is that moving objects get shorter in the direction of motion.

Although you can't actually see it, you can detect the effects of this on a wire carrying a current. Electricity is essentially electrons moving from one atom to another along the wire. As they leave one atom, it momentarily becomes positively charged because it has lost a negatively-charged electron. It's then referred to as an 'ion'.

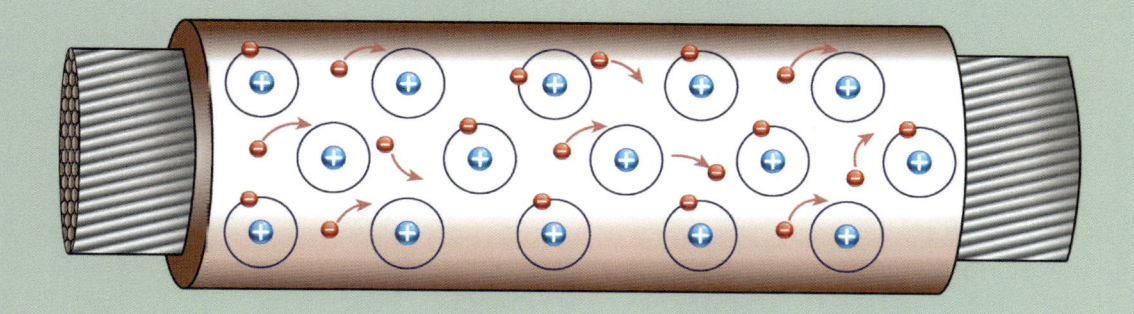

So a stream of electrons is constantly moving along the wire and because they are moving you observe a relativistic contraction – they get closer together. Consequently, from your point of view there are more free electrons than ions in a length of wire and the imbalance generates a magnetic field surrounding the wire. The contraction is extremely small, but because there are so many electrons (there are more atoms in a single inch of standard-gauge copper wire than there are grains of sand in the world) and because the electrostatic force between protons and electrons is so high, a strong field will result.

If it wasn't for the magnetic field effect, dynamos wouldn't work and everything would have to be battery-powered.

THE MATHS

If you can remember Pythagoras' theorem and simple algebra from your schooldays you can not only understand time slowing and length contraction but you can work out by how much they change at different speeds. Showing that $E = mc^2$ requires more advanced knowledge, but is still within the reach of a high school mathematician.

If you don't feel you need those details, then you can skip this chapter. The important thing is to have understood the concepts of Special Relativity, which I hope you now do.

Time dilation

Let's first take the slowing of time, or time dilation as it's formally called.

Going back to the switching on of a light in a railway carriage, we'll say that the train is travelling at a speed of *v* and light at a speed of *c*. (The units don't matter – feet per second, miles per hour, etc. So long as you keep everything in the same units it works out)

We'll also say that for a person on the train the light appears to go directly to the floor, taking t_v seconds. For a person on the platform, the light takes a longer, diagonal route and takes t_0 seconds.

Now, speed is distance divided by time, so by simple algebra, distance = speed x time. Thus the distance travelled by the photons is ct_v for the passenger but ct_0 along the diagonal for the observer. Also, the distance the train travels as viewed from the platform while the light moves from source to floor is vt_0.

We can draw a right-angled triangle to show all this.

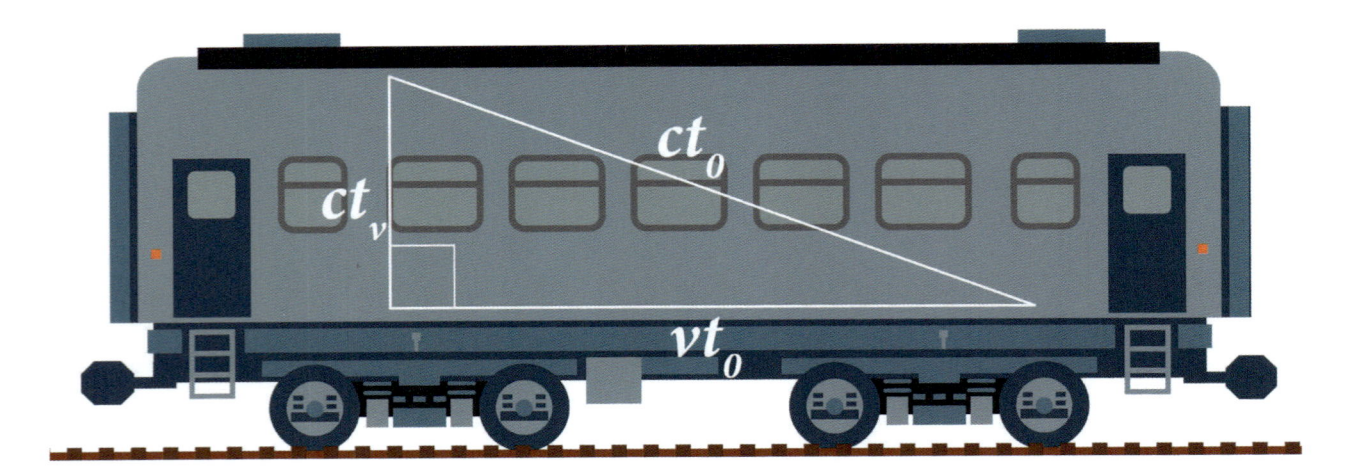

Now we can apply Pythagoras' theorem: the square on the hypotenuse is the sum of the squares on the other two sides, so: $(ct_0)^2 = (ct_v)^2 + (vt_0)^2$.

It works out more neatly if we recognise that time dilation works both ways – to those on the train, events happen more slowly on the platform. So we can equally well say that: $(ct_v)^2 = (ct_0)^2 + (vt_v)^2$
Expanding and rearranging we have: $c^2 t_0^2 = t_v^2 (c^2 - v^2)$

Dividing both sides by c^2 gives, $t_0^2 = t_v^2 (1 - \frac{v^2}{c^2})$

Finally, rearranging and taking the square root of both sides gives, $t_v = \frac{t_0}{\sqrt{1 - \frac{v^2}{c^2}}}$

Now we have a formula for time on a moving frame of reference (t_v) compared with time on our own frame (t_0).

Let's use this to work out a simple example. If a spaceship is travelling at three-quarters of the speed of light, how much slower does time pass in the ship from our point of view?

It's easier here to work out the $(\frac{v^2}{c^2})$ bit first. This is $\frac{(0.75c)^2}{c^2}$ which as c^2 cancels, equals $0.75^2 = 0.5625$.

So the final result is $\frac{t_0}{\sqrt{1 - 0.5625}}$ which you can work out on your calculator is a little more than $1.5 t_0$. The answer then is that at three quarters of the speed of light, from the point of view of an earth observer, for every second that ticks by, one-and-a-half seconds appear to pass on the spaceship.

The Lorentz Transformation

The expression $\frac{1}{\sqrt{1-\frac{v^2}{c^2}}}$ occurs regularly in relativistic calculations. We just used it for time dilation and it will also appear in calculating length contraction and in proving the equivalence of mass and energy.

This expression is called the 'Lorentz transformation' after the Dutch mathematician Henrik Lorentz. He derived it in an effort to explain the conclusions of James Clerk Maxwell.

Because it appears so often, physicists often replace the expression with the Greek letter *gamma* (γ) as it saves space and looks impressive.

The point to notice is that the speed of light is so immense, that with almost any velocity we encounter $\frac{v^2}{c^2}$ is very close to zero, and therefore the Lorentz transformation is almost exactly equivalent to multiplying by one.

Looking at the graph below, we can see that until you get to about 30% of the speed of light (say 100,000 miles per second) relativistic effects would be almost imperceptible.

Length contraction

We can use the time dilation formula to conveniently work out the length contraction in the direction of motion.

Going back to our train, let's start by drawing a marked length on the platform.

As the train passes, we measure the length of time it takes for any point on the train to travel the length of our marker. We have two observers for this: one on the train and one on the platform.

Let's say that the person on the platform observes the time taken by a train moving at velocity v as t_0 and the person on the train observes t_v.

As before, distance is speed × time, so the observer on the platform calculates the marker length $l_0 = v\, t_0$ and the observer on the train finds the length $l_v = v\, t_v$.

We can equate to eliminate v, and rearrange to find: $l_v = \dfrac{l_0 t_v}{t_0}$.

We earlier derived the formula for time dilation as:

$t_v = \dfrac{t_0}{\sqrt{1-\dfrac{v^2}{c^2}}}$ but as from the point of view of a passenger time moves more slowly on the platform, we can equally well say that: $t_0 = \dfrac{t_v}{\sqrt{1-\dfrac{v^2}{c^2}}}$

Substituting for t_0 in the previous equation we get: $l_v = \dfrac{l_0 t_v \sqrt{1-\dfrac{v^2}{t^2}}}{t_v}$

Now we simply cancel t_v to give the final formula of: $l_v = l_0 \sqrt{1-\dfrac{v^2}{c^2}}$

This gives us the contracted length on the train (l_v) equivalent to the marked length on the platform (l_0).

Again, we'll apply this to a spaceship travelling at three-quarters of the speed of light.

As before, $\left(\dfrac{v^2}{c^2}\right)$ is 0.5625 and therefore

$l_v = l_0 \sqrt{1 - 0.5625}$

Using your calculator, you find that a metre measured on the spaceship appears to be about 0.66 metres when observed from earth.

Mass increase

When we looked at the children in the train tossing a tennis ball to each other, we said that the energy in a moving object is made up of its mass and its velocity. Multiplying these two properties together gives us a quantity known as *momentum*.

The total momentum cannot change, so we can say that $m_0 v_0 = m_v v_v$ where m_0 is the mass of the ball observed from the platform and m_v the mass observed in the train, and similarly v_0 is the velocity of the ball observed from the platform and v_v the velocity observed in the train.

As velocity is distance divided by time, this becomes $m_0 \dfrac{s}{t_0} = m_v \dfrac{s}{t_v}$

We earlier derived the equation for time dilation as: $t_v = \dfrac{t_0}{\sqrt{1 - \dfrac{v^2}{c^2}}}$

Substituting for t_v in the first equation gives $m_0 \dfrac{s}{t_0} = \dfrac{m_v s \sqrt{1 - \dfrac{v^2}{c^2}}}{t_0}$

We can now cancel s and t_0 and rearrange to give: $m_v = \dfrac{m_0}{\sqrt{1 - \dfrac{v^2}{c^2}}}$

Now we have a formula for the observed mass on a moving frame of reference (m_v) compared with mass on our own frame (m_0).

Galileo

Galileo was the first person to systematically study and quantify the laws of motion. As it was very difficult to observe or time a freely falling weight in those days, he reasoned that a ball rolling down a grooved inclined plane would be equivalent in many ways.

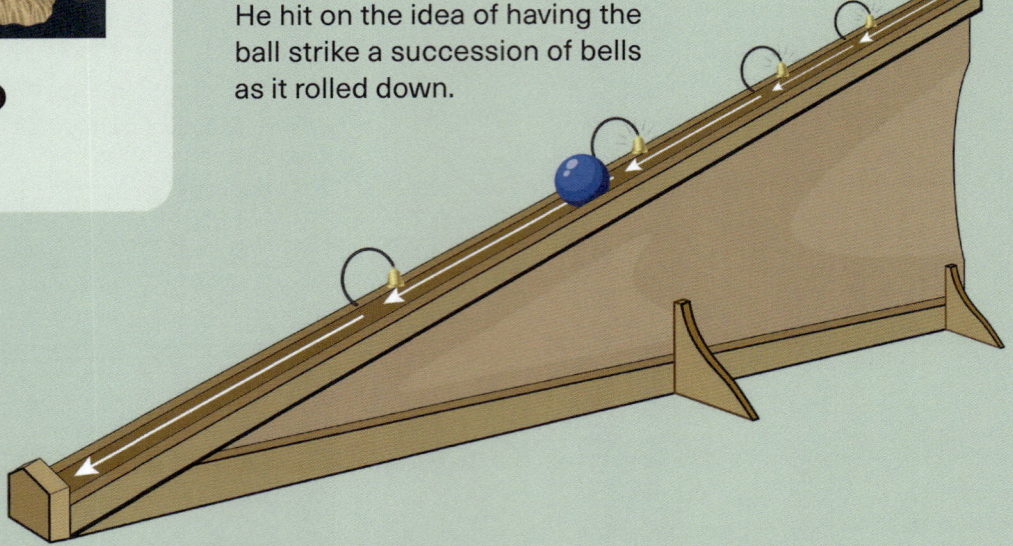

He hit on the idea of having the ball strike a succession of bells as it rolled down.

He adjusted the positions of the bells so that as the ball accelerated he heard a succession of regular chimes. (His father was a musician which may have given him the idea)

The human ear is sensitive to a regular beat – most people can detect an error of as little as 1/64th of a second – so by this means he was able to deduce the principles of acceleration under gravity.

Building on Galileo's experimental insights, we can now derive the relationship between distance, acceleration, and time.

The average speed over a given distance in a given time is $\frac{s}{t}$. But average speed can also be expressed as the starting speed plus the final speed divided by two which is $\frac{(v+v_0)}{2}$

Equating these and rearranging we get: $s = \frac{(v+v_0)}{2} t$

The definition of acceleration is the change in velocity over time, that is:
$a = \frac{(v-v_0)}{t}$ so $v = v_0 + at$

Substituting for v in our last equation, $s = \frac{(v_0 + at + v_0)}{2} t$

And therefore $s = v_0 t + \frac{1}{2} a t^2$

In cases where motion is from rest, we have the useful formula for the distance travelled in a given time as: $s = \frac{1}{2} a t^2$

The equivalence of mass and energy

To set the scene, we need to understand the concepts of "force", which is an impulse that acts on a body and "work" which is the accumulated energy needed to maintain that force over a distance.

Imagine a rocket ship in space.

If it fires its engine for a short while, it will accelerate to a certain speed and retain that speed indefinitely in the absence of air resistance or anything else to slow it down. This is Newton's first law: an object remains at rest or in uniform motion in a straight line unless acted on by an external force.

But suppose it continues to fire its engine? In that case it will keep gathering speed – it will accelerate. The amount of force exerted at any point is defined as the mass of the spaceship multiplied by the acceleration that results, that is: $F = m\,a$. When doing calculations, we most commonly talk of "newtons" of force; one newton being the force needed to give a kilogram of mass an acceleration of one metre per second per second.

As the spaceship continues to accelerate, we can see that it is gaining ENERGY – the faster it goes, the more effort it would take to bring it to rest. The amount of energy gained in a given distance of travel is the force multiplied by the distance: $Work = F\,s$. We measure work in "joules" which is defined as the work required to exert a force of one newton over a distance of one metre.

Energy can exist in a variety of forms and often one form can be changed into another (although the total amount of energy can never change). In the case of our spaceship, the work done increases "kinetic energy". Kinetic energy is the energy stored in a moving object by virtue of its mass and its speed. So in this case, work done and kinetic energy gained are equivalent.

We can start with the useful equation $s = \frac{1}{2}at^2$ where s is the distance travelled for given acceleration and time. (You may remember this from school, but if not it's derived in the box on Galileo). By definition, $v = at$ and so $t = \frac{v}{a}$. Equating and rearranging we have $s = \frac{v^2}{2a}$

We showed above that work done is defined as force times distance and force is mass times acceleration, so: Work = $F s = m a s$

Using the formula for s above we have: Work $= m a \frac{v^2}{2a} = \frac{1}{2} m v^2$

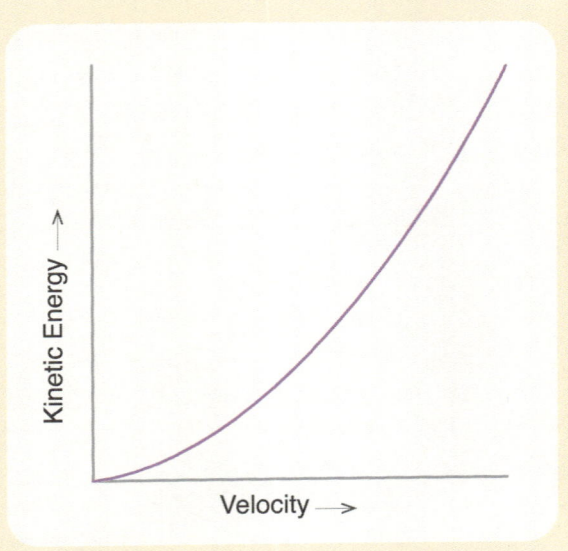

The work done imparts energy to a body, therefore the formula for the kinetic energy gained in reaching velocity v is:
$KE = \frac{1}{2} m v^2$

We can plot this as a graph, and we can see that in classical mechanics, where mass is a constant and there is no upper limit to velocity, kinetic energy can increase indefinitely.

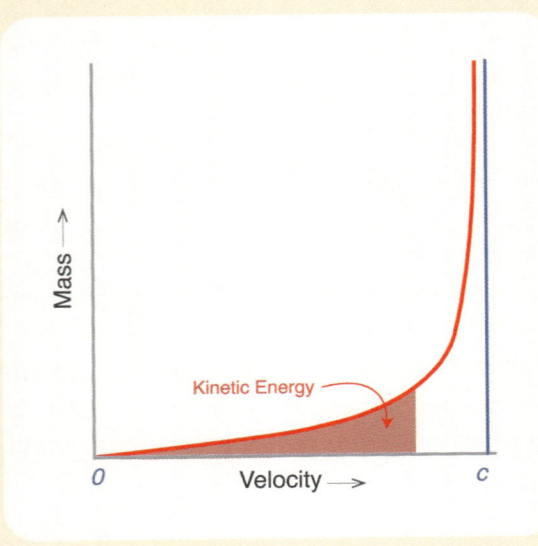

But as we've seen, mass increases as the object moves faster relative to us and the velocity can never exceed the speed of light, so we need to plot a new graph of mass against velocity.

Now to find the amount of kinetic energy accumulated in reaching a given speed, we can no longer simply read it off but must measure the shaded area under the curve.

Measuring that area can be done in various ways. We can for example get a good approximation by replacing the curve with narrow vertical bars which can be easily calculated individually, then added up. Of course, the narrower the bars, the more accurate the result, but to do the measurement precisely takes us into an advanced mathematical technique called calculus.

Calculus uses a clever trick that causes the bars to be of zero width and therefore the result to be completely accurate. If you're not familiar with calculus, skip to the end of the chapter where I've derived an equation giving the area.

We can write an expression in integral form for the kinetic energy acquired over a given velocity range, allowing for the fact that mass varies with velocity, as:

$$KE = \int_0^v v \, d(mv)$$

As both m and v vary, we have to 'integrate by parts' using the general rule which can stated as:

$$\int a \, db = ab - \int b \, da$$

In our case, we'll set $a = v$ and $db = d(m\,v)$

This means that $da = dv$ and $b = [m\,v]$

Applying the rule we have: $KE = v\Big[mv\Big]_0^v - \int_0^v mv \, dv$

Which simplifies to: $KE = m\,v^2 - \int_0^v mv \, dv$

Now we can substitute the relativistic expression for mass we derived earlier. Given a 'rest mass' of m_0 the 'relativistic mass' m_v increases with velocity as:

$$m_v = \frac{m_0}{\sqrt{1 - \frac{v^2}{c^2}}}$$

We substitute this in the second term, giving:

$$KE = mv^2 - m_0 \int_0^v \frac{v \, dv}{\sqrt{1 - \frac{v^2}{c^2}}}$$

This is a rather tricky integral, but we can approach it as follows:

Firstly, let $u = 1 - \frac{v^2}{c^2}$

Then we differentiate u to give: $du = -\frac{2v}{c^2} dv$

And rearrange to give: $dv = \frac{-c^2}{2v} du$

We can now replace dv with du in the original equation to give:

$$KE = mv^2 - m_0 \int_0^v \frac{v}{\sqrt{u}} \left(-\frac{c^2 \, du}{2v}\right)$$

We move the constants outside the integral and cancel *v*. We also recall that raising to the power of ½ is equivalent to a square root and a negative exponent is equivalent to the reciprocal, giving:

$$KE = m v^2 + \frac{m_0 c^2}{2} \int_0^v u^{-½} \, du$$

We now perform the integration to give:

$$KE = mv^2 + \frac{m_0 c^2}{2} \left[2\sqrt{u}\right]_0^v$$

We cancel the 2's and substitute back for u to give the equation:

$$KE = m v^2 + m_0 c^2 \left[\sqrt{1 - \frac{v^2}{c^2}}\right]_0^v$$

Since the limit of *v* is *c* (the speed of light), we can let *v = c*. Then evaluating the integral, the first term goes to zero and the second to unity, giving:

$$KE = m c^2 - m_0 c^2$$

This is a remarkable result. It shows that the kinetic energy of a body does not directly depend on its velocity, but rather on the relativistic increase in mass, $(m - m_0)$, resulting from its motion.

We can also see, that when a body is at rest and kinetic energy can be ignored, it still possesses an energy component $m_0 c^2$ resulting from its 'rest mass' which can famously be written as:

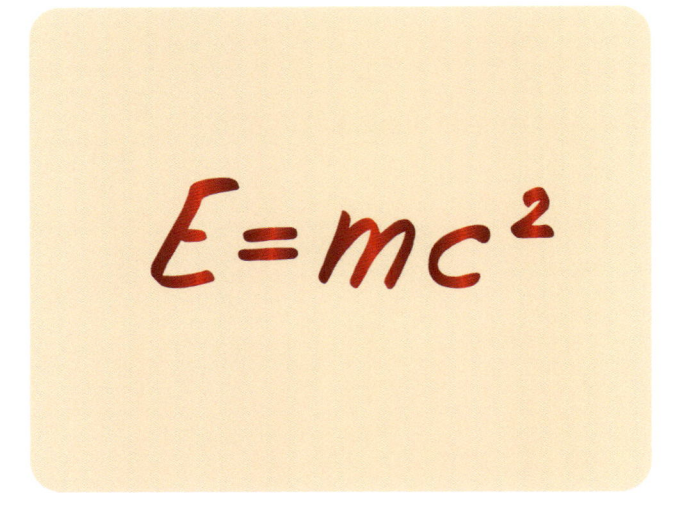

$$E = mc^2$$

Printed in Dunstable, United Kingdom